BASIC PRINCIPLES AND FOUNDATIONS OF BIOSAFETY AND LABORATORY SECURITY IN A TUBERCULOSIS AND HIV/AIDS REFERENCE LABORATORY.

Philip Anochie

BASIC PRINCIPLES AND FOUNDATIONS OF BIOSAFETY AND LABORATORY SECURITY IN THE TUBERCULOSIS AND HIV/AIDS REFERENCE LABORATORY .

Philip Anochie

BASIC PRINCIPLES AND FOUNDATIONS OF BIOSAFETY AND LABORATORY IN THE TUBERCULOSIS AND HIV/AIDS REFERENCE LABORATORY .

PRINCIPLES AND FOUNDATIONS OF SAFETY PRACTICES AND STANDARDS IN A TUBERCULOSIS AND HIV/AIDS REFERENCE LABORATORY.

Philip Anochie

BASIC PRINCIPLES AND FOUNDATIONS OF BIOSAFETY AND LABORATORY SECURITY IN A

TUBERCULOSIS AND HIV/AIDS REFERENCE LABORATORY.

Published by

Philip Anochie

Philip Nelson Institute of Medical Research, Lagos, Nigeria.

BASIC PRINCIPLES AND FOUNDATIONS OF BIOSAFETY AND LABORATORY SECURITY IN A

TUBERCULOSIS AND HIV/AIDS REFERENCE LABORATORY.

Published by

Philip Anochie

Philip Nelson Institute of Medical Research, Lagos, Nigeria.

Your comments and suggestions about this book would be appreciated. Please submit them to Curriculum Development. Philip Nelson Institute of Medical Research, 11, John Fayemi Close, P.O. BOX 55601, Falomo, Ikoyi, Lagos, 101008, Nigeria. E-mail: philipnelsoninstitute@yahoo.com.

Your comments and suggestions about this handbook would be appreciated. Please submit them to Curriculum Development. Philip Nelson Institute of Medical Research, 11, John Fayemi Close, P.O. BOX 55601, Falomo, Ikoyi, Lagos, 101008, Nigeria. E-mail: philipnelsoninstitute@yahoo.com.

Please list your name, address, e-mail and telephone number. Be sure to give the title of the book. Then offer your comments and suggestions about the book's strengths and areas of potential improvement.

First published:

ISBN:

©

Office of the Publisher

Philip Nelson Institute of Medical Research, Lagos, Nigeria.

Printed in USA.   English approval:

CONTENTS

Cha RINCIPLES AND FOUNDATIONS OF BIOSAFETY AND LABORATORY SECURITY IN A TUBERCULOSIS AND HIV/AIDS REFERENCE LABORATORY.

Philip Anochie

BASIC PRINCIPLES AND FOUNDATIONS OF BIOSAFETY AND LABORATORY SECURITY IN THE TUBERCULOSIS AND HIV/AIDS REFERENCE LABORATORY .

Philip Anochie

BASIC PRINCIPLES AND FOUNDATIONS OF BIOSAFETY AND LABORATORY IN THE TUBERCULOSIS AND HIV/AIDS REFERENCE LABORATORY .

PRINCIPLES AND FOUNDATIONS OF SAFETY PRACTICES AND STANDARDS IN A TUBERCULOSIS AND HIV/AIDS REFERENCE LABORATORY.

Philip Anochie

BASIC PRINCIPLES AND FOUNDATIONS OF BIOSAFETY AND LABORATORY SECURITY IN A

TUBERCULOSIS AND HIV/AIDS REFERENCE LABORATORY.

Published by

Philip Anochie

Philip Nelson Institute of Medical Research, Lagos, Nigeria.

BASIC PRINCIPLES AND FOUNDATIONS OF BIOSAFETY AND LABORATORY SECURITY IN A

TUBERCULOSIS AND HIV/AIDS REFERENCE LABORATORY.

Published by

Philip Anochie

Philip Nelson Institute of Medical Research, Lagos, Nigeria.

Your comments and suggestions about this book would be appreciated. Please submit them to Curriculum Development. Philip Nelson Institute of Medical Research, 11, John Fayemi Close, P.O. BOX 55601, Falomo, Ikoyi, Lagos, 101008, Nigeria. E-mail: philipnelsoninstitute@yahoo.com.

Your comments and suggestions about this handbook would be appreciated. Please submit them to Curriculum Development. Philip Nelson Institute of Medical Research, 11, John Fayemi Close, P.O. BOX 55601, Falomo, Ikoyi, Lagos, 101008, Nigeria. E-mail: philipnelsoninstitute@yahoo.com.

Please list your name, address, e-mail and telephone number. Be sure to give the title of the book. Then offer your comments and suggestions about the book's strengths and areas of potential improvement.

First published:

ISBN:

©

Office of the Publisher

Philip Nelson Institute of Medical Research, Lagos, Nigeria.

Printed in USA.   English approval:

CONTENTS

SUMMARY

CHAPTER 14 IMPLICATIONS OF POOR LABORATORY SAFETY AND SECURITY PRACTICES.

CONCLUSION

BIBLIOGRAPHY

SUMMARY

This book examines the biological safety principles, characterization of four biosafety levels, areas of research needs , safety management programme profile, key tuberculosis and HIV/AIDS reference laboratory security protocols and the socio-economic implications of poor safety practices in a tuberculosis and HIV/AIDS reference laboratory.

Laboratory safety programmes are plans for preventing sickness and injury to personnel and damage or destruction of physical assets. The fundamental objectives of a meaningful laboratory safety programme include improvement of safety skills and attitude of all personnel , and development of a surveillance programme for prompt identification of hazards.

It also explains fully, the term "containment" which describes safe methods for managing infectious materials in the laboratory environment where they are being handled or maintained.

The purpose of containment is to reduce or eliminate exposure of TB/HIV/AIDS reference laboratory workers, other persons and the outside environment to potentially hazardous agents.

Primary containment, the protection of personnel and the immediate laboratory environment from exposure to infectious agents , is provided by both good microbiological techniques and the use of appropriate safety equipment to ensure biological safety and security in the tuberculosis and HIV/AIDS reference laboratories.

CHAPTER 1

INTRODUCTION

The use of vaccines may provide an increasing level of personnel protection. Secondary containment , the protection of the environment external to the Tuberculosis (TB) , Human immune deficiency virus (HIV), and Acquired immune deficiency syndrome (AIDS) reference laboratory from exposure to infectious materials, is provided by a combination of facility design and operational practice.

Therefore, the three elements of containment include laboratory practice and technique, safety equipment and facility design and also construction of secondary barriers.

The risk assessment of the work to be done with a specific agent will determine the appropriate combination of these elements.

CHAPTER 2

BIOSAFETY LEVELS IN A TUBERCULOSIS AND HIV/AIDS REFERENCE LABORATORY.

Four biosafety levels (BSLs) are described in biomedical literature which consists of combinations of laboratory facilities. Each combination is specifically appropriate for the operations performed, and the laboratory function or activity.

The Laboratory Director is specifically and primarily responsible for assessing the risks and appropriately applying the recommended biosafety levels.

CHAPTER 3

BIOSAFETY LEVEL 1 PRACTICES.

Safety equipment and facility design as well as construction are appropriate for undergraduate and secondary educational training and teaching laboratories and for other laboratories in which work is done with defined and characterized strains of viable microorganisms not known to consistently cause diseases in healthy adult humans.

Bacillus subtilis, Naegleria gruberi , infectious canine hepatitis and exempt organisms under the NIH Recombinant DNA guidelines are representative of microorganisms meeting this criteria.

Biosafety level 1(P1) represents a basic level of containment that relies on standard microbiological practices with no special primary or secondary barriers recommended other than a sink for hand washing.

CHAPTER 4

BIOSAFETY LEVEL II PRACTICES.

Equipment, facility design and construction are applicable to clinical diagnostic , teaching and other laboratories in which work is done with the broad spectrum of indigenous moderate – risk agents that are present in the community and associated with human disease of varying severity.

With good microbiological techniques, these agents can be used safely in activities conducted on the open bench, provided the potential for producing splashes or aerosols is low. Hepatitis B Virus (HBV), HIV , the *Salmonellae* and *Toxoplasma* species are representative of microorganisms assigned to this containment level.

Biosafety level III (P2) is appropriate when work is done with any human- derived blood, body fluids, tissues or primary human cell lines where the presence of an infectious agent may be unknown e.g; Hepatitis and other highly infectious viruses, fungi and bacteria, TB/HIV/AIDS reference laboratories.

Tuberculosis and HIV/AIDS reference laboratory personnel working with human – derived materials should refer to the OSHA Blood borne pathogen standards for specific required precautions [1]

Primary hazards to personnel working with these agents include accidental percutaneous or mucus membrane exposures or ingestion of infectious materials. Extreme caution should be taken with contaminated needles or sharp instruments.

Even though organisms routinely manipulated at Biosafety level II are not known to be transmissible by aerosol route, procedures with aerosol or high splash potential that may increase the risk of such personnel exposure must be conducted in primary containment equipment or in devices such as biological safety or safety centrifuge cups.

Other primary barriers should be used as appropriate such as splash shields, face protection, gowns and gloves. Secondary barriers such as hand washing sinks and waste contamination facilities must be available to reduce potential environmental contamination.

CHAPTER 5

BIOSAFETY LEVEL III PRACTICES.

Equipment, facility design and construction are applicable to clinical , diagnostic , teaching, research or production facilities in which work is done with indigenous or exotic agents with a potential for respiratory transmission and which may cause serious and potentially lethal infection.

Mycobacterium tuberculosis ,that cause tuberculosis (TB), *St. Louis encephalitis virus* and *Coxiella burnetti* are representative of the microorganisms assigned to this level.

Primary hazards to personnel working with these agents relate to autoinoculation , ingestion and exposure to infectious aerosols in a TB laboratory.

CHAPTER 6

BIOSAFETY LEVEL III PRACTICES.

At Biosafety level III (P3) , more emphasis is placed on primary and secondary barriers to protect personnel in contagious areas, the community and the environment from potentially infectious aerosols. For example; all laboratory manipulations should be performed in a biological safety cabinet (BSC) or other enclosed equipment such as a gas-tight aerosol generation chamber.

Secondary barriers for this level include controlled access to the laboratory and ventilation requirements that minimize the release of infectious aerosols from the TB/HIV/AIDS reference laboratory.

CHAPTER 7

BIOSAFETY LEVEL IV PRACTICES.

Safety equipment, facility design and construction are applicable for work with more dangerous , exotic , foreign and strange agents that pose a high individual risk life – threatening diseases, which may be transmitted via the aerosol route ie. Ebola virus, SARS , influenza, Marburg and Monkey pox virus as well as multidrug and extensive drug resistant tuberculosis , and for which there is no available vaccine or therapy.

Agents with a close or identical antigenic relationship to biosafety level IV practice agents should also be handled at this level. When sufficient data are obtained, work with these agents may continue at this level or at a lower level.

Viruses such as Marburg or Congo – Crimean haemorrhagic fever or SARS are manipulated at biosafety level IV practice. The primary hazards to personnel working with Biological safety level IV agents are respiratory exposure to infectious aerosols , mucus membrane or broken skin exposure to infectious diagnostic materials, isolates and naturally or experimentally infected animals pose a high risk of exposure and infection to laboratory personnel, the community and the environment.

The laboratory worker's complete isolation from aerosolized infectious materials is accomplished primarily by working in a class III BSC or a full- body , air – supplied positive- pressure personnel suit.

The BSL 4 facility itself is a generally a separate building or completely isolated zone with complex, specialized ventilation requirement and waste management systems to prevent release of viable agents to the environment.

The Laboratory Director is specifically and primarily responsible for the operation of the laboratory. His/her knowledge and judgments are critical in assessing risks and appropriately represents those conditions under which the agents can ordinarily be safely handled.

Special characteristics of the agents used, the training and experience of personnel and the nature or function of the laboratory may further influence the director in applying these recommendations.

TB/HIV/AIDS reference laboratories and other clinical laboratories , especially those in health care facilities, receive clinical specimens with requests for a variety of diagnostic and clinical support services.

Typically, the infectious nature of clinical material is unknown and specimens are often submitted with a broad request for microbiological examination for multiple agents (e.g. sputa submitted for " routine" , acid – fast and fungal cultures) . It is the responsibility of the laboratory director to establish standard procedures in the laboratory, which realistically address the issue of the infective hazard for the director to establish standard procedures in the laboratory , which will realistically address the clinical specimens.

Except in ordinary circumstances, (e.g. suspected haemorrhagic fever) the initial processing of clinical specimens and serological identification of isolate can be done safely at BSL 2, the recommended level for work with blood borne pathogens such as HBV and HIV.

The containment element described in BSL IV (P4) are consistent with OSHA standard " Occupational safety and health administration (OSHA)). This requires the use of specific precautions with all clinical specimens of blood or other potentially infectious material (Universal or standard precaution) .

Additionally, other recommendations specific for clinical laboratories may be obtained from the American National Committee for Clinical Laboratory Standards (NCCLS).

The segregation of clinical laboratory functions and limited or restricted access to such areas is the responsibility of the laboratory director. It is also the Director's responsibility to establish standard written procedures that address the potential hazards and the required precautions to be implemented.

Selection of an appropriate biosafety level for work with a particular agent or animal study depends upon a number of factors.

Some of the most important are the virulence , pathogenicity, biological stability, route of spread and communicability of the agent, the nature or function of the laboratory, the procedures and manipulations involving the agent , the endemicity of the agent and the availability of effective vaccines or therapeutic measures.

CHAPTER 8

LABORATORY SECURITY IN A TB/HIV/AIDS REFERENCE LABORATORY.

This is a very important aspect of laboratory management. The following guidelines were developed by the Centers for Disease Control and Prevention Atlanta, USA (CDC) [1] and adapted Vanderbilt Environmental Health and Safety (VEHS) [2] to address laboratory security for laboratories using biological agents or toxins capable of causing serious or fatal illness to humans or animals.

The guidelines are reflected below in the next chapters.

CHAPTER 9

RECOGNIZE THAT LABORATORY SECURITY IS RELATED BUT DIFFERENT TO LABORATORY SAFETY.

This is done by involving both safety and security experts in evaluation and development of recommendation for a given facility or laboratory.

There should be review of safety and security procedures regularly. Management should review policies to ensure that they have good policies and procedures for both laboratory staff and visitors. Laboratory supervisors should ensure that both laboratory workers and visitors understand the requirements and are trained and equipped to follow established procedures. Review and safety and security policies and procedures whenever an incident occurs or a new threat is identified should be regularly implemented.

CHAPTER 19

CONTROL ACCESS TO AREAS WHERE BIOLOGIC AGENT OR TOXINS ARE USED AND STORED.

Laboratories and animal care areas should be separate from the public areas of the buildings in which they are located. Laboratory and animal care areas should be locked at all times. Card keys or similar keys should be used as permit to laboratory animal areas .

All entries (including entries by visitors , maintenance workers , repairmen and others needing one-time or occasional entry) should be recorded either by the card key device (preferable) or by signature in a log book.

Only workers required to perform a job should be allowed in laboratory areas and workers should be allowed only in areas and at hours required to perform their particular job. Access for students , visiting scientists etc. should be limited to hours when regular employees are present. Access for routine cleaning , maintenance and repairs should be limited to hours when regular employees are present.

Freezers, refrigerators, cabinets and other containers where stocks of biological agents, hazardous chemicals or radioactive materials are stored should be locked when they are not in direct view (eg. When located in unattended storage areas).

CHAPTER 20

KNOW WHO IS IN THE LABORATORY AREAS.

Facility administrators and laboratory directors should know all workers. Depending on the biological agents involved and the type of work being done , a background check and /or security clearance may be appropriate before new employees are assigned to the laboratory area.

All workers (including students , visiting scientists and other short-term workers) should wear visible identification badges and escorted or cleared for entry using the same procedures as for regular workers.

CHAPTER 21

KNOW WHAT MATERIALS ARE BEING BROUGHT INTO THE LABORATORY AREA.

All packages should be screened (Visual or X-ray) before being brought into the laboratory area. Packages containing specimens, bacterial or virus isolates or toxins should be opened in a safety cabinet or other appropriate containment device. Know what materials are being removed from the laboratory area. Biological materials and toxins for shipment to other laboratories should be packaged and labeled in conformance with all applicable local , federal and international shipping regulations.

Required permits should be in hand before materials are prepared for shipment. The recipient (preferably) or receiving facility should be known to sender and the sender should make an effort to ensure that materials are shipped to a facility equipped to handle those materials safely.

Hand carrying of microbiological materials and toxins to other facilities is rarely appropriate. If biological material or toxins are to be hand carried on common carrier, all applicable regulation must be followed.

Containment or possibly contaminated materials should be decontaminated before they leave the laboratory area. Chemicals and radioactive materials should be disposed of in accordance with local , state and federal regulations.

CHAPTER 22

HAVE AN EMERGENCY PLAN.

Control of access to laboratory areas can make an emergency response more difficult. This must be considered when emergency plans are developed. An evaluation of the laboratory area by appropriate facility personnel with outside experts if necessary to identify both safety and security concerns to be conducted before an emergency plan is developed.

Facility administrators, laboratory directors, principal investigators, laboratory workers , the facility safety office and facility security officials should be involved in emergency planning. Police, fire and emergency responders should be informed as to the types of biological materials in use in the laboratory areas and assisted in planning their responses to emergencies in the area.

Plan should include provision for immediate notification of of (and response by) laboratory directors, laboratory workers, safety office personnel or other knowledgeable individual when an emergency occurs, so that they can deal with biosafety issues if they occur.

Laboratory emergency planning should be coordinated with facility-wide plans. Such facts like bomb threats, severe weather (hurricanes , floods) , earthquakes, power outages, and other natural or unnatural disasters should be considered when developing laboratory emergency plans.

CHAPTER 23

HAVE A PROTOCOL FOR REPORTING INCIDENTS.

Laboratory directors, in cooperation with facility safety and security officials should have policies and procedures in place for reporting and investigation of incidents or possible incidents (e.g. undocumented visitors, missing chemicals, unusual or threatening phone calls etc.).

IMPLICATIONS OF POOR LABORATORY SAFETY AND SECURITY PRACTICES.

Laboratories are unique environments in relation to the safety of those who work within them. Clinical specimens received from patients pose a great deal of hazard to laboratory personnel because of the infectious agents that they may contain.

In addition to the threat of infection, laboratories also contain safety risks associated with any institutional environment namely ; those of fire, electrical and chemical hazards, hazardous environmental situations (slippery floors, faulty air-handling systems), radioactive materials, equipment malfunction and dangers imposed by natural disasters.

Laboratory safety programmes are plans for preventing sickness and injury to personnel and damage or destruction of physical assets. The fundamental objectives of a meaningful laboratory safety programme include improvement of safety skills and attitude of all personnel , and development of a surveillance programme for prompt identification of hazards. Others are formulation of plans for prompt correction of all hazards and co-ordination of laboratory safety effort with the overall laboratory safety and security programme ie. An employee health programme that must include pre- employment physical examination , with laboratory and radiological studies to establish fitness for laboratory employment.

This should be periodically repeated and employees should report all work-related illnesses and accidents. A general safety programme must include the orientation of new employees to department's attitude and policies for assuring safe laboratory conduct such as orderly housekeeping standards; storage and arrangement of supplies, eating , drinking, smoking and safe attire, safety officer and coordinated efforts for isolation of communicable diseases, control of nosocomial infections and plans for dealing with fire and other disasters.

A programme for handling chemicals which must include policies and labeling , transporting , storage, dispensing, and disposing of all chemicals. A programme for handling of all biological specimens which must include instructions for collecting , transferring, storage, processing and disposing of all specimens and specimen containers as well as instructions for hand washing and the care of work surfaces.

A fire prevention programme must include physical facilities and operational practices that satisfy fire code; handling and storing of combustibles, instructions for operating all heat- generating equipment (gas burners) ; and well –conceived and rehearsed plans in the event of fire e.g; sand buckets, fire extinguishers and fire blankets.

A first aid programme must include policies for dealing with all job-related injuries.

The health output of an individual leads to various social and economic consequences for that individual, the household, and the community as reflected in such measures as income, savings, physical and human capital, productivity , health consumption and social interaction.

These consequences are produced through a variety of response mechanism depending upon the severity of the disease and clinical manifestation. The difficulty of defining disease is implied in the very structure of the world "disease" . So many different kinds of disturbances can make a person feel not at ease and lead him to seek the aid of a physician that the word ought to encompass most of the difficulties inherent in the human conditions.

Generally, among the lay public, disease implies some serious organic and psychic malady such as cancer or insanity. Modern medicine in practice is broadening this concept to refer to any state, organic or psychic , real or imaginary , that disturbs a person's sense of well- being. In this sense , disease may threaten life or simply interfere with its enjoyment. It may prevent the sick person from functioning as a normal human being or simply from reaching his self-selected goals.

It is now realized that in dealing with the problems of the " disease" person, subjective and social factors may be as important as the objective organic lesions of behavioral disturbances recognized by the physician or psychiatrist.

Human history and our personal lives continue to be shaped by epidemics, ie. sudden outbreaks of infectious disease within a community . Many of the factors that determine the occurrence of disease in an individual also influence the spread of infectious diseases throughout a population.

Environmental conditions that bring organisms and hosts together encourage the spread of endemic disease. These environmental conditions include any physical, chemical, biological or social factors that are essential to the survival and transmission of the infectious agents. Only when these major elements- an infectious agent, a susceptible host and proper environmental conditions work in concert does infectious disease emerge and persist in a community.

The laboratory as an environment poses many hazards to unsuspecting and untrained people especially visitors. Epidemiology is the study of disease distribution. Social and cultural factors are important determinants of disease aetiology and distribution through their influence on the relationship between a human population and its natural environment, or through their direct influence on the relationship between a human population and its natural environment, or through their direct influence on the health of the population.

Social and cultural distinctions associated with differences in age, sex, occupation, class, ethnicity and community can have significant effects.

The incidence of numerous acute infections is highest in childhood, indicating that as people grow older, they develop immunities that decrease their vulnerability to these diseases. Death rates are clearly related to age. Obviously, these epidemiological patterns reflect biological variations in vulnerability to sickness and death associated with age differences .

It is therefore essential that good safety practices are employed to protect children and elderly relatives of laboratory personnel from laboratory- acquired infections.

The picture with respect to morbidity differences between the sexes is complete. But if male and female role distinctions can influence differences in response to illness, they can also influence differences in the development of illness as well; particularly if the culture emphasizes such role distinctions ; for example: house-keeping staff in the laboratories in Africa are usually females.

Studies of the effects of occupation on diseases have been an important part of epidemiology and has indicated that susceptibility to disease varied in accordance with means of livelihood.

In the laboratory, the unanticipated hazards of handling newly discovered, poorly understood, or previously un-encountered microbes exist alongside its myriads of problems. For example, the accidental infection of 31 researchers in Marburg , Germany in 1967 by a previously unknown virus in the tissues of African green monkeys is a remarkable event.

A substantial part of epidemiological research has been devoted to the influence of social stratification and ethnic differences on disease prevalence and aetiology.

This influence can be particularly significant in nutritional maladies and in certain infectious diseases whose spread is affected by the material conditions of life. For example, when people are crowded together indoors, respiratory infections (like common cold in schools , offices and meningococcal infections in military recruits) spread rapidly.

This is perhaps why respiratory infections are common in winter. The air in ill-ventilated rooms is also more humid , favouring survival of suspended microorganisms such as *Streptococci* and enveloped viruses.

A laboratory staff or visitor with laboratory associated infection may serve as an effective vehicle of transmission. Association of disease frequency with contrasting community setting , have formed another focus of epidemiological interest. Social correlates of rural – urban distinctions and their implications for health have been significant problems.

The role of patient, especially in many seriously disabling illnesses , may be extremely difficult. Even in so many cases, if a patient recovers or temporarily overcomes the effects of a deadly disease , he may not only regain his status in the community, but, even become an object of admiration. However, some diseases , in addition to actual or supposed impairment of the individual , carry with them an onerous burden of sigma; a social definition of disease that transforms the victim into a social outcast.

The mode of infection, does not alter the stigmatization, therefore, no one will admit having HIV infection , even if laboratory associated.

The prevention of illness and containment of disease are part of every medical system. But much more is involved than questions of sanitation, private and public cleanliness and robustness, for notions of contagion are bound up with religion and world view, and with perceptions of the powers and intentions of one's families , neighbors and friends; not to mention strangers.

Though intents may not be subject to control or change, behavior to some extent is , and therefore the medical system especially with regard to contagion and sanitation , is directly hooked into local systems of social organization and control.

Medical phenomena can be indicative of the performance of a social system. The health of its population is one significant test of effectiveness with which a society functions [3] .

One way to stay healthy and to insure the well-being of one's family , neighbors, friends and colleagues is to strictly observe all laboratory safety and security considerations. It is recommended that access to TB/HIV/AIDS laboratories should be limited to those who have been informed on how they can perform safely and securely.

Visitors , especially small children should be discouraged. Certain areas of TB/HIV/AIDS reference laboratories should be closed to visitors.

Finally, healthcare providers are expected to provide employees with all devices and mechanisms called engineering controls , necessary to protect them from hazards encountered during the course of work and in cases of accidents, immunization and first aid.

CONCLUSION

The last several documented cases of small pox in the world, were acquired by employees working in a different area in the building where a small pox research laboratory was located. The principal Investigator was so dismayed that his possible deficient safety practices led to these small pox cases that he committed suicide.

It is the legal responsibility of the laboratory director and supervisor to ensure that an exposure control plan has been implemented and that the mandated safety guidelines are followed. Scientific exploration is not without hazards.

Each year , these explorers contract infectious diseases because of accidental exposure to pathogens in laboratories. Over 4,000 cases of laboratory associated infections have been recorded with more than 160 fatalities. The centers for disease control and prevention has defined which microbes currently represent the greatest hazard in the laboratory and has classified etiologic agents into four categories according to biosafety class levels.

Class 1 contains agents that pose little risk of serious disease. These include *Staphyllococcus epidermidis* and many other members of the normal flora.

Microbes in class 4 are the most dangerous and require the highest degree of containment. The plague bacillus is a class 4 –bacterium.

The greatest number of laboratory infections , has occurred among persons engaged in research activity.

Fewer than 23% of these infections have been reported as occurring in TB/HIV/AIDS diagnostic laboratories. This perhaps reflects the unanticipated hazard of handling newly undiscovered , poorly understood or previously un-encountered microbes.

For example, in 1967, 31 persons in Hamburg , Germany , as mentioned above, handling the tissues of African green monkeys were infected by a previously unknown virus in the tissue, and 6 of the victims died.

Most laboratory- associated infections are acquired by contact with infectious aerosols i.e. in the case of tuberculosis . Random air sampling demonstrates that common laboratory manipulations release microorganisms into the atmosphere e.g. blowing out pipette, removing stoppers or centrifugation.

Infected experimental animals or man may discharge contaminated respiratory droplets into the air, in addition, bites and scratches from these animals occasionally cause infections.

Mouth pipetting is one of the most hazardous of all laboratory manipulations, resulting in numerous cases of typhoid fever, tularemia, scarlet fever, hepatitis and influenza as well as other bacterial, parasitic and fungal opportunistic infections of HIV/AIDS.

Accidental inoculation with needles and syringes accounts for many laboratory associated infections such as HIV.

Spills of infectious materials are also implicated. In cases where a highly virulent pathogen is introduced into a susceptible population, an epidemic may occur. The affected population may be as small as a single family or as large as the global community.

BIBLIOGRAPHY:

1. CDC office of Health and Safety (1999). Principles of Biosafety . BMBL Section II. In: http://www.cdc.gov/od/ohs/biosfty/bmb14/bm.

2. Vanderbilt Environmental Health and Safety (2002). Laboratory Security. In: http://www.safety.vanderbiltedu/resources/biosafetysecurity.htm.

3. McKane , L. and Kandel. (1996). Principles of Epidemiology. Microbiology: *Essentials and Applications*. Mc Graw Hill Inc. (Pub): 501-523.

www.ingramcontent.com/pod-product-compliance
Lightning Source LLC
Chambersburg PA
CBHW021852170526
45157CB00006B/2408